MW00476818

HANDBOOK FOR VOLUNTEERS OF THE IRISH REPUBLICAN ARMY

NOTES ON GUERRILLA WARFARE

PALADIN PRESS
BOULDER, COLORADO

*Handbook for Volunteers of the Irish Republican Army:
Notes on Guerrilla Warfare*

Copyright © 1985 by Paladin Press

ISBN 13: 978-1-58160-570-9
Printed in the United States of America

Published by Paladin Press, a division of
Paladin Enterprises, Inc.
Gunbarrel Tech Center
7077 Winchester Circle
Boulder, Colorado 80301 USA
+1.303.443.7250

Direct inquiries and/or orders to the above address.

PALADIN, PALADIN PRESS, and the "horse head" design
are trademarks belonging to Paladin Enterprises and
registered in United States Patent and Trademark Office.

Visit our Web site at www.paladin-press.com

CONTENTS

A Handbook for Volunteers of the Irish Republican Army,
issued by General Headquarters, 1956.

CHAPTER 1 – OUR TRADITION

No nation has a greater tradition of guerrilla warfare than Ireland. Our history is full of examples of its successful use. We have produced some fine guerrilla leaders whose true qualities have never been fully assessed.

Their strength lay in the support they received from the Irish people. In the final analysis it was the people who bore the enemy's reprisals. Whoever betrayed the cause, or gave up the fight, or suffered loss of spirit, it was seldom the people.

KERNE OF OLD

The kerne of old were lightly armed foot soldiers. Their tactics were of the skirmishing kind. They harassed the Normans. In open or positional warfare they had no hope of breaching the defences of the strongly-armed, iron-clad Normans.

Art Og MacMorrough Kavanagh was a typical guerrilla leader of his period. Richard II of England came twice with large armies to subdue him (1394 and 1399) and never succeeded. Another was Fiach MacHugh O'Byrne and yet a third Leinster leader was Rory O'More. The O'Byrne's great victory at Glenmalure followed the strict application of guerrilla tactics.

Ulster produced its quota of which Shane (the Proud) O'Neill was only one. The English leader Sydney paid him a fine tribute when he said: "He armeth and weaponeth all the peasants of his country, the first that ever did so of an Irishman."

But it was left to the finest military leader Irish history has produced, Hugh (the Great) O'Neill, to understand fully the potentialities of guerrilla warfare. He proved it too for 9 years.

YELLOW FORD

O'Neill forsake his guerrilla tactics only once: at Kinsale. He did so under pressure from the Spaniards who had landed at the wrong place and time and who were insistent that he attack the encircling English army of Mountjoy. This was the first and last battle in which he adopted open positional warfare. The result we all know.

How he defeated Essex, who landed here with huge armies in an attempt to subdue him, and how these armies were dissipated in vain attempts to reach him, is worthy of study by the serious student of guerrilla tactics. He knew when to strike and when to withdraw and he never fought a battle except on his own terms. The battle of the Yellow Ford is the greatest example of this.

At the Yellow Ford he lured Marshal Bagenal with 5,000 men out of Armagh. He had full intelligence as to Bagenal's strength and battle order. He hit him with snipers all the way to the Blackwater and at a spot called the Yellow Ford—in a prepared position—he gave battle, split Bagenal's superior forces, and destroyed them division by division. Bagenal was killed, his army's retreat became a rout, three-fourths of the English forces were annihilated. It was, in effect, a large scale ambush executed in the classical Cannae manner: by drawing the enemy into the centre and closing him in a pincer.

OWEN ROE

Another O'Neill, Owen Roe (a regular officer of the Spanish army), was also a master guerrilla tactician. His defeat of General Munroe (June 1646) at Benburb is an example of how a small, well trained guerrilla force can destroy a far superior army.

First, Owen Roe's cavalry cut off Munroe's reinforcements. Then his cavalry returned and attacked the enemy's big guns. Then they swept back the unguarded foot soldiers. The victory was complete: 5,000 poorly-armed men against a standing army of 6,000. A memorable victory in any country's story.

1798's GUERRILLAS

Michael Dwyer is a much neglected figure in Irish history. It is too easily forgotten that he held out in the Wicklow mountains for seven years against England's forces. His guerrilla force grew from 10 men who came with him after the disastrous Battle of Tara to more than 150. He used every tactic in the guerrilla handbook and was never really defeated.

But Michael Dwyer's example is important in other ways. He could not hope to win as long as only his small force was in action. Had other groups risen similarly throughout the country, the outcome would have been far different. But the true application of guerrilla tactics to a revolutionary situation was not properly understood at the time.

FINTAN LALOR

James Fintan Lalor (1848 leader) never got the opportunity to carry his guerrilla theories into the battlefield. But that he understood these tactics is evident from his writings. Take this quotation:

"The force of England is entrenched and fortified. You must draw it out of position; break up its mass; break its trained line of march and manoeuvre, its equal step and serried array . . . nullify its tactic and strategy, as well as its discipline; decompose the science and system of war, and resolve them into their first elements."

FENIANS

Lalor, after his 1849 abortive rebellion, fathered the Fenian

3

movement through such figures as John O'Mahony and James Stephens. But one Fenian figure above all practised guerrilla warfare extensively and again like Michael Dwyer failed because his tactics were little understood and he was left isolated. This man was Captain Mackey Lomasey, affectionately known in the Cork area where he operated as "The Little Captain," an Irish-American who harried English garrisons for arms for a long time before being taken. Lomasey learned his guerrilla tactics while an officer with the Union forces in the American Civil War.

BLACK AND TANS

By guerrilla warfare after 1916, with the united resistance of the Irish people to British rule almost a fact, and spearheaded by the I.R.A., it became quite obvious that England could no longer govern Ireland. By the time of the Treaty it has been calculated England could not have reconquered Ireland with less than 100,000 troops aided by all the accoutrements of modern warfare.

It is now almost an accepted historical fact that had Lloyd George's bluff been called during the Treaty negotiations—and that he was bluffing is no longer in doubt—the outcome would have been a free Ireland. The bluff was, of course, that he would declare "immediate and terrible war."

But guerrilla operations which made this great success possible had to have a united people behind them. British Government in Ireland no longer existed in fact. British terror in Ireland could not hope to revive it. And terror had come as a last resort but one. The final one was to split the people.

The hammer blows of the guerrillas destroyed the British administration. The guerrillas acted in small numbers in the right localities and compelled the British to disperse to find them. Then as they searched they hit them at will by means of the ambush. Communications were systematically destroyed

4

and even the British army's transport system in the country was disorganised. The enemy's intelligence service was completely dislocated. The R.I.C.—the eyes and ears of British rule—was demoralised. British justice courts could not operate—for the people ignored them. The British gradually were forced to evacuate the smaller more isolated garrisons. They concentrated in the larger towns. The areas evacuated came under sole control of The Republic. The next step was to isolate the larger centres and keep cutting communications and constantly hitting the enemy. In time these would have been evacuated too. Thus ended the last great phase of guerrilla operations against British rule in Ireland.

CHAPTER 2—WHAT IS GUERRILLA WARFARE?

A small nation fighting for freedom can only hope to defeat an oppressor or occupying power by means of guerrilla warfare. The enemy's superiority in manpower, resources, materials, and everything else that goes into the waging of successful war can only be overcome by the correct application of guerrilla methods.

Guerrilla warfare might be defined as the resistance of all the people to enemy power. In the struggle the guerrillas act as the spearhead of the resistance.

Up to the second world war the military textbooks ignored this phase of warfare. After that they couldn't afford to ignore it. Now the General Staffs are working out methods of dealing with guerrillas. Britain has built an independent Brigade to deal with them. In the age of the H-Bomb, strangely enough, the tactics of guerrillas are being widely copied.

For example, the former British Chief of the General Staff

and Commander of the 14th Army, Field Marshal Sir William Slim, has this to say on the matter:

"Dispersed fighting, whether the dispersal is caused by the terrain, the lack of supplies or by the weapons of the enemy, will have two main requirements—skilled and determined junior leaders and self-reliant, physically-hard, well disciplined troops.

"Success in future land operations will depend on the immediate availability of such leaders and such soldiers, ready to operate in small independent formations. They will have to be prepared to do without regular lines of communications, to guide themselves and to subsist largely on what the country offers.

"Unseen, unheard and unsuspected, they will converge on the enemy, and, when they do reveal themselves in strength, they will be so close to him that he will be unable to atomise them without destroying himself."

That then is the blueprint of warfare in the atomic age—the tactics and strategy of guerrillas.

REGULAR WARFARE

In regular warfare the tactical objective is to destroy the enemy in battle by concentrating superior numbers at a decisive time and place. The guerrilla strikes not one large blow but many little ones; he hits suddenly, gnaws at the enemy's strength, achieves surprise, disengages himself, withdraws, disperses and hits again.

A regular army unit depends on all kinds of tactical support: air, ground, communication, supplies, armour, artillery, reserves, units to left, right and rear. And so on. There are all kinds of weapons available. Plans are worked out by General Staffs, transmitted through a chain of command down the line. Attacks will go in under artillery, air and even sea barrages. Armour will create the breakthrough.

More often than not the line soldiers are unaware of what is

6

happening or is supposed to happen. They rely on N.C.O.s and officers in all eventualities. They are trained to fight as cogs in an intricate and vast machine embracing perhaps millions like themselves.

THE GUERRILLA

The situation of a guerrilla is quite different.

Outside of the support he gets from the people among whom he operates—and this support must never be underestimated for it is vital to his eventual success—he fights alone. He is part of an independent formation that is in effect an army by itself. He must be **self-contained**.

If necessary he must act alone and fight alone with the weapons at his disposal—and these very often will not be of the best. He must find his own supplies. His endurance has to be great: and for this he needs a fit body and an alert mind. Above all he must know what he is fighting for—and why.

The guerrilla must move fast and hit hard. He must be adaptable. He must change his methods constantly.

His training must be such that during withdrawal his formation can break up and reform later. It is not his job to hold a line or take a city or maintain a strategically vital area.

But what he must do is this:

He must exhaust the enemy by constant harassment.

He must attack constantly and from all directions.

He must stage successful retreats, return to the attack, avoid encounters with the enemy that are not of his own making.

Tactics have to be changed constantly. Formations have to be independent of terrain and lines of communications. This is what is meant by being self-contained.

The guerrilla never affords the enemy a target. He is bold in the attack and his great advantage is **mobility**.

The plan of action must be simple, understood by all, and—if

7

possible—well rehearsed. The guerrilla's great weapon is **surprise**. To achieve this surprise, intelligence must be first-class. The guerrilla must know everything about the enemy and his battle-order, his strength and his weakness—even his plans for anti-guerrilla activities.

Good intelligence breeds good morale. And for the guerrilla morale is everything. It is this morale that gives the guerrilla his determination and his daring.

Once the fight is joined, it must be carried out relentlessly and to the bitter end. The road may be long, the sacrifices great, but if the guerrilla has this endurance and the will to win, he cannot be defeated.

To strive constantly towards these goals day by day, mounting small successes on bigger victories, building up the morale of the people, these are the aims of the volunteer guerrilla. They ensure final victory.

CHAPTER 3—GUERRILLA STRATEGY

The strategy of guerrilla warfare is to build up resistance centres throughout the occupied area and confine the enemy to the larger towns by restricting his movements and communications. In time the resistance centres are knitted together into one liberated area.

After that the job is to drive him out of his supposedly safe base: and thus out of the country.

The essence of all strategy is to bring, by the use of surprise and mobility—or a combination of both—the greatest possible strength to bear at a chosen time and place. It must be ensured that the enemy does not—or is not able to—assemble **his** strength at that point.

This holds true also of guerrilla warfare. But it involves clever manoeuvre and here the skill of the commander, the organisation of his forces and his mobility, play an important role. The guerrilla attempts to do three things:

(1) Drain the enemy's manpower and resources.

(2) Lead the resistance of the people to enemy occupation.

(3) Break down the enemy's administration.

He achieves the **first** by the very fact of his existence and his constant harassment of the enemy. He remembers that his own task is not to hold ground but to ensure that in time the enemy will not hold any either.

He achieves the **second** by remembering that the people will bear the brunt of the enemy's reprisal tactics and by inspiring them with aims of the movement. In this way they will be made tenacious and strong for in the long run it is the people who can stop the enemy: by their backing of the national movement.

And he achieves **three** when the enemy imposes martial law and thus recognises he can no longer rule that area in the old way. In effect he is recognising that the people no longer want him.

As the enemy recognises the new phase he has entered he makes ever greater attempts to destroy the guerrillas. And the guerrillas' number one task is to ensure that he fails.

The main principles of war can be boiled down to these five:

Concentration—or economy of force.

Protection—to guard against being taken unawares by the enemy.

Surprise—to catch the enemy in such a situation that he is unable to switch his forces to meet the attack. Surprise can be of time or place as well as armament and forces.

Aggressiveness—determination to knock out the enemy in the attack.

Objective—to pursue the objective to the end despite the enemy's counter-measures and never to be sidetracked.

These principles hold good for guerrilla warfare also.

CHAPTER 4–TACTICS OF GUERRILLAS

The guerrilla is always on the offensive. To be successful he needs surprise, mobility, exact knowledge of the enemy, determination, fire action and shock action. And these, as in warfare generally, will be determined by his armament, his training, his morale, his planning, and co-ordination between higher command and the initiative of the guerrilla detachment.

He destroys enemy communications, raids enemy bases, ambushes the enemy when the latter tries to find him. But his tactics must always be in process of change because in time the enemy will have an answer for the old ones.

In guerrilla warfare the attacks must be fluid and carried out by detachments or columns, sections and even battle teams. The local commander will have the initiative in his hands.

The battle teams may move by bounds to the objective and if attacked will be capable of dissolving immediately—only to come together again later. To reach the objective the battle teams may have to infiltrate hostile strong points. This would be done under cover of darkness. The infiltration would take place through lanes previously reconnoitred.

INITIATIVE

Since the guerrilla's main task is to drain the enemy it follows that he hits the enemy in his most vulnerable area and fights for the initiative—never ground.

The guerrilla will not fight the enemy in a long battle where reserves would overwhelm him: he strikes only when he can win. And he avoids superior forces. When the enemy advances, he withdraws. When the enemy rests, he hits him. He attacks when the enemy is exhausted. And when the enemy counterattacks, the guerrilla flees.

THE GUERRILLA NEVER AFFORDS A TARGET.

10

The guerrilla relies on **surprise** above all the principles of war. He makes as little contact as possible with the enemy and uses "the smallest force in the quickest time at the farthest place" (Lawrence).

His attacks are planned and rehearsed. He strikes when the enemy is moving, resting or is lightly guarded. By using flank attacks on the line of march a much weaker force can achieve success.

THE DECISION

The guerrilla must be the master of lightning blows to achieve fast decisions. When conditions are unfavourable he disperses and shifts base immediately. He will do that when:

(1) He hasn't the forces to hold off the enemy.

(2) He is encircled and has to break out.

(3) The terrain is unfavourable.

(4) His supplies are insufficient.

Guerrilla units will only concentrate their force when the enemy is advancing and can be damaged severely. Otherwise they remain intact in small detachments. This reduces errors and makes action easier.

Also the small detachment working independently can make better use of ground and is able to break-out of an enemy cordon more easily.

The guerrilla must pick his targets wisely. His choice must lie with the ones that play up his natural advantages and which aid his general strategy.

DECEPTION

The guerrilla must muster the tactics of deceiving the enemy. He moves secretly and quickly. He travels light, is highly mobile, knows the terrain intimately and uses it to its best advantage. He uses ground, darkness and fog to aid his attacks. He may move by night and sleep by day.

11

He sticks to back routes, avoids main roads and towns, seldom uses motor transport, gets to his objective by forced marches, changes direction constantly. Closely guarded territory he crosses by battle teams (two men).

His movements will be by bounds and well guarded to front, rear and flanks. The detachment will then reassemble at a predetermined point near area of attack. The guerrilla must be a master of the manoeuvre.

In deceiving the enemy as to his methods and intentions the guerrilla will use many ruses. He can always cover his tracks by leaving fast-moving snipers and smaller formations scattered over wide areas while the main forces pull out.

CONDITIONS

But a guerrilla movement can only operate if two conditions exist. They are:

(1) That there are guerrilla formations everywhere (centres of resistance) and that they are everywhere in action. Thus when the enemy concentrates on one point another blazes up. Without this he could snuff out the guerrillas in no time.

(2) That the guerrilla detachments are SELF-CONTAINED in everything needed for their operations including arms, supplies, intelligence and propaganda among the people.

Co-operation of the people is also vital to the guerrillas. Because it has to be stressed that support for the aims of the guerrillas must come from the population. Cut loose from the people, a guerrilla formation can neither develop nor survive.

And every guerrilla formation is an educator of the people. It exposes the lies of the enemy, shows the reasons for his occupation.

LARGE-SCALE

Guerrilla operations generally begin on a small level and constantly grow. By the time centres of resistance have been built and tied together large guerrilla formations may carry out combined operations. Here the use of radio communication is invaluable. Certain formations, with a high degree of mobility, will be used to intervene at decisive points. In such cases the guerrilla formations will have beforehand the general plan of action. They will be able to act with a high degree of independence, depending on the situation, as a result.

But the guerrilla even at this point must be careful about moving on to positional warfare. He has to remember that his main task is to keep the enemy off balance. He will win if he does this.

CHAPTER 5—ORGANISATION AND ARMS

The organisation of a guerrilla force in the field must in no way duplicate that of a regular army. Its objective, tasks, need and outlook are completely different. Its structure must be elastic so that it can fit in with terrain conditions and operational necessity.

Three points must be made clear at the outset:

(1) The membership of a guerrilla force operating in a particular area should in the main consist of locals. Since small blunders may lead to major setbacks the guerrillas should know the terrain like a book. For this knowledge locals are invaluable.

(2) The guerrillas are volunteers and are inspired by an ideal. Therefore their loyalty, understanding of what is at stake and

13

discipline will be—and must be—on a much higher level than that obtaining in a regular army.

(3) Leadership will not come so much by appointment as by the trust the guerrillas place in their commander. He must be worthy of that trust if he is to succeed.

From these points follow a number of principles:

(a) Organisation will vary according to the conditions. Above all it must not be rigid.

(b) Instead of discipline of the regular army type there will be a more stern battle discipline: agreement on the job to be done, and the need to do it, and obedience to the guerrilla code, these take the place of the unthinking army type discipline.

(c) Breaches of the guerrilla code—desertion, betrayal, breach of confidence in any way—must be severely dealt with on the spot.

(d) Guerrillas' work in decentralised or dispersed units. The independent detachment (15 to 25 men) is the key to the organisational structure. The detachment will decide its own local targets and carry out its job within an area without further orders. It can expect little help if it fails in its mission. It takes its punishment alone.

Accordingly organisational co-ordination is not important. But there may be, as the occasion demands, operational co-ordination of several guerrilla detachments.

(e) Since guerrillas are self-contained they dispense with the supply and reserve methods of regular formations—where one fighting man may be backed up by as many as 10 non-combatants.

THE COLUMN

From this it follows that organisationally the basic guerrilla unit is the independent detachment—or as we in Ireland named it, the Flying Column. Its strength will follow development and local needs. Operationally, it is under higher command but at the same time knows its own field of operations and carries out its tasks without further checking.

It may be called on by higher command to carry out certain tasks or support other columns in the field. But most of its time will be taken up with local operations. The column should be able to live and fight on its own for a long period of time ... without help from any quarter. And it does the following:

(1) Picks its own targets except when acting under direct orders of higher command.

(2) Co-ordinates its activities with other columns through higher command.

(3) Gets its reserves and replacements from the local population.

(4) Is responsible for its own security, intelligence, arms, equipment, supplies, and propaganda among the people.

(5) To operate outside its allotted territory it must get the sanction of higher command. Also it passes along all intelligence data collected to higher command.

SECTIONS

The flying column should be as small as is operationally possible. It should seldom exceed 30 men. When it grows stronger new columns may be created. Its size aids its fast manoeuvrability.

The basis of all tactics is fire and movement. The column will concentrate more firepower for less men and thus achieve striking superiority over enemy forces when in the attack.

It would break down in this fashion:

(1) **The battle team**—consisting of two men for fire and movement. The team could be a tactical entity when required: one would give fire support while the other manoeuvred.

(2) **The section**—consists of two battle teams and the section commander—five men. Three sections make up the basic

15

column. The commanders and column commander are the column H.Q. and every volunteer is a fighting man. The battle teams and sections will be trained to operate on their own.

ARMS

Almost any small arms weapon can be a guerrilla weapon. For assault: sub-machine-guns, light machine-guns, shotguns, explosives, grenades, pistols, automatic rifles, flame throwers are needed. For support, light machine-guns, rifles, 2" mortars (for high angled fire) and rocket launchers.

But it will be the fortunate column that can have all or most of these weapons. For the most part the guerrillas have to make and improvise.

For many guerrilla operations explosives can be used to make up arms deficiencies. Entrances can be made to buildings by the use of surprisingly small charges. (Always on the building's blind side.)

Breaking down enemy resistance is also easier once explosives are employed. Charges of one pound H.E. can be used to mop up the enemy. And pole charges (10 pounds or more of H.E.) from the end of a long pole fired by a time fuse are invaluable for pillboxes and higher windows and such like.

Charges can also be used in ambushes. Buried along the approach route (9 inches to a foot in depth) in a quantity of four pounds to a foot they can be fired electrically from cover. The enemy trucks, once stopped, can then be attacked from nearby, also using explosives.

Four-pound charges can also be spaced for guerrilla protection in much the same manner as anti-tank mines. As well, explosives can be strapped to ten-foot planks for use against enemy trucks and vehicles and other points.

The employment of explosives, once their tactical use is

understood, can have many variations. They are only dangerous in the hands of inexperienced men—once the usual precautions have been taken. Fire is also a potent guerrilla weapon.

The guerrilla armament must be light and carry concentrated firepower for shock engagements. Heavier weapons may be used in the fire base while lighter and closer ranged ones are employed in the final assault.

Every man in the column should be able to handle all weapons of the column, maintain them and repair them.

Task of the column: To get as near as possible unobserved to the target and then use surprise and shock tactics to win.

CHAPTER 6—WITH THE PEOPLE

Successful guerrilla operations involve the people. It is the quality of their resistance to the enemy and support for the guerrillas which in the end will be the decisive factor. The guerrillas are the spearhead of the people's resistance.

In fact, a guerrilla force will be unable to operate in an area where the people are hostile to its aims. And it must be remembered always that it is the people who will bear the brunt of the enemy's retaliatory measures.

Accordingly there should be constant contact and co-ordination between the guerrillas and the local population. This is aided by:

(1) Recruiting volunteers for columns from population of territory in which column is operating.

(2) Use of civil political committees among the people whose function it would be to agitate against the oppressor, get new members for guerrillas, organise supplies for columns, provide transportation for guerrillas, lead the people in a campaign of active and passive resistance to enemy occupation.

17

(3) Have guerrilla agents working among civilian population collecting information for army.

(4) Use of part-time guerrillas who would continue in civilian occupations yet be available for active service when called on. Thus local companies would be built up and used as reserves when the occasion warranted.

(5) Build up liaison between guerrillas and people until such time as it was perfect. When the people suffer under enemy oppression for aiding the guerrillas the latter would help and protect the people.

INFORMATION

To build up the resistance of the people to the required pitch needs more than guerrilla activity. The aims of the movement must be popularised, the objectives clearly stated, and the world must be informed of what is happening—and why.

This type of information is actually good education too. Part of the education process is countering the enemy's propaganda. The basic idea is that the guerrilla education campaign must be continuous, must beat the enemy at his propaganda game and must expose his lies to the people and indeed to the whole world.

This end of the guerrilla operations is no less important than the destruction of enemy resources and bases.

Information must be factual to build up confidence among the people in the national movement. What it must do is this:

(1) Give the people tenacity to stand up to the enemy by showing them the struggle is worthwhile and necessary.

They must be made aware that the national struggle will be victorious in the end—but that the end depends on them.

(2) Get world public opinion behind the just fight of the people.

(3) Undermine the enemy's morale and his propaganda by

exposing his methods and by constant emphasis on the unjustness of his cause.

(4) Be the spiritual mainspring of those actively engaged in the national movement so that they understand the need to destroy the enemy and his power forever.

METHODS

The main channels of information available to the guerrillas are newspapers, leaflets, radio, word of mouth. Other methods may be worked out and new ones invented. For example: Painting of slogans, proclamations and manifestoes and so on.

All the means of winning the confidence of the people must be utilised. The ideas of the movement must be so popularised that no one is in doubt—least of all the enemy—that it will win eventually.

This information service must function continuously to get maximum results. Among the things it must do are:

(1) Show weakness of enemy position and propaganda used to bolster that position.

(2) Show what is wrong with political and social order.

(3) Suggest remedies and how they can be brought about.

(4) Be in touch all the times with thinking of the people.

ACCURACY

The guerrillas' information service will be judged on its accuracy. It must tell the people exactly what is required of them. It must show that the guerrilla movement is all-knowing, all-powerful, a part of the people's life—the people are a part of it.

Its broadcasting must be so interesting that friend and foe alike tune in. This rule applies to all information media.

19

The information service of the guerrilla command should work in close liaison with intelligence so that information is up to the minute and accurate and yet does not give away information to the enemy. It would also independently collect and evaluate and distribute its own information. The world has to be informed of what is happening for it will be basic enemy policy to shut off all contact between the movement and the rest of the world.

The world must know and understand what is being done, what the enemy is trying to destroy and why, and the way these things can be ended and peace restored and freedom won.

The use of regular bulletins for foreign newspapers and news-agencies becomes a necessity. The bulletin should be of the documentary type: no room for emotional pleas or the like. Just the facts.

CHAPTER 7–GUERRILLA BASES

The guerrilla column will have a base to serve as a rallying point, as an assembly area after an attack or a withdrawal, for use as a training camp, a place for the wounded, and for general regrouping.

From these bases the enemy will be raided, his communications cut, his strongpoints subjected to heavy and consistent attacks. Outlying enemy areas, adjacent to the bases, must be destroyed beforehand.

The bases can be large or small, elaborate or quite simple. If the guerrilla movement is sufficiently strong and well developed the bases will reflect this and may indeed be quite large. Otherwise they should be scattered, inconspicuous and well hidden.

At best the bases can only be semi-permanent. They will be

changed as often as the situation warrants. Large ones suffer the disadvantage of attracting enemy attention: they offer him a target. And, of course, the whole aim of his strategy is to be able to do just that.

Guerrilla bases must have a good line of withdrawal in case of attack. Indeed they should have several routes of escape. If possible there should be only one entrance and the base should be located in an inaccessible area—mountains, marshes, uninhabited places.

They should be changed frequently.

DEFENCE

The defence of a guerrilla base must be so organised that:

(1) A few snipers (acting as look-outs) can cover approaches for long distances. There should be a clear-cut system of alarms and a well-worked drill for evacuation of volunteers and supplies.

(2) There should be an emergency exit so that the attackers can be hit from the rear.

(3) The lay-out of the column dug-outs should be such that all sections of the column are in a position to manoeuvre or completely dissolve as the occasion warrants. Best is the triangular form. Each dug-out would give supporting fire to the other.

(4) Dug-outs should be camouflaged for defence from air and ground attack, have an all-round traverse and be well camouflaged.

(5) There should be caches for arms and supplies which would be insulated against water and drainage.

(6) Approaches may be mined and bases near main roads should have exits facing away from the road.

(7) After the capture of guerrillas belonging to the column the bases must immediately be changed. A force might, how-

ever, be left behind to surprise the enemy if he attacks.

POINTS TO NOTE

In guarding bases sentries should work in pairs and have frequent relief. They should have definite instructions on the line to take in case of attack or if a civilian blunders on the base.

In the case of innocent blundering on the hide-out, the sentries should be able to grab the individual and leave subsequent matters to the commander. In the case of an attack, the drill for the occasion should go into operation without delay. The base must never be surprised.

Camouflage is vital to the digging of any dug-out. The line should fade into the background. If caves are used the opening might be covered by dirtied sacking curtained off by leaves and twigs. But foliage must be changed often—(in case of withering) and tracks must be carefully covered for they show from the air.

Finally, bases are for the storing of arms and for training while resting. They are not places for lying low and the guerrilla must avoid the temptation to use them as such.

CHAPTER 8–GUERRILLA ATTACK

The regular soldier is no match for the trained guerrilla in attack. Because the guerrilla holds the initiative, strikes when he is ready, uses shock action and surprise to attain his ends, then breaks contact and withdraws.

Guerrilla attacks are fluid. There is a large measure of decentralisation, and manoeuvres are carried out by battle teams who know their job and do it.

The battle teams move by bounds to the objective and may

have to infiltrate hostile strong points. They cloak themselves in darkness. They infiltrate through previously reconnoitred lanes. When they hit the strongpoint they use their first waves to cut enemy communications, support and supplies. Then the final subjection is accomplished by the use of close fire-power—small arms, LMGs, mortars, high explosives, thermit and flame throwers. Or whatever weapons the guerrilla commander has at his disposal.

The guerrilla turns night into day. He knows that at night accurate aimed fire is impossible. That control is very difficult even for seasoned troops. That a force can hit and then dissolve. He exploits these advantages and becomes adept at night operations.

SIMPLICITY

Certain rules govern all attack situations. Thorough reconnaissance is necessary before the assault. The enemy's strength and defences are ascertained. The guerrillas move by night and rest by day. They get as near as possible without being seen.

The guerrilla takes it for granted that his movements may come to the enemy's attention so he guards against surprise. He avoids the local population. His movements must give no indication of his eventual target.

He will, of course, be self-contained. His equipment will be as light as possible. He will be well camouflaged, wear light clothes, tape grenades and other equipment that might rattle. He darkens his hands and face—burnt cork is good for this.

With accurate information on the enemy's defences the guerrilla commander may stage a rehearsal. Every man will know his job. Every man is fully briefed on his role and the plan itself.

The plan must be simple. Complicated manoeuvres always break down. The attack must be timed with precision and it should move at top speed. There must be total surprise, a thorough carrying out of the job to be done, and then a planned withdrawal.

WITHDRAWAL

The guerrilla must live to fight another day and for that reason the withdrawal is as important as the attack itself. There should be no slip-up. Assembly areas must be clearly understood and an alternative rendezvous should be named in case of interference with the original plan.

There must also be a plan for evacuation of the wounded.

And the withdrawal routes must be well chosen so that there is no confusion.

After the attack a full conference should be held by the guerrillas. Every aspect of the job will be examined. Mistakes will be pointed out. Weaknesses and strength of the enemy will be gauged.

More than any other soldier the fighting morale of the guerrilla must be on a very high plane. Every volunteer is imbued with aggressive confidence in his fighting skill and the importance of that skill to his comrades and to his people. This pride will lead him to do the apparently **impossible**—and the "impossible" will seem easy as a consequence. His enthusiasm will be infectious and generate such power that no force on earth can stop it.

For the fight he is engaged in is worthy of nothing less.

CHAPTER 9—ENEMY TACTICS

The enemy will use every device he can command to destroy the guerrilla movement. These will include: martial law, curfews, blocking of roads leading to towns, cities and villages and his own garrisons, round-ups, night patrols, use of hostages, reprisals, propaganda to alienate local population from the guerrillas, torture of prisoners . . .

Tactically he will attempt any or all of the following to destroy the guerrilla columns:

(1) **Encirclement**—He throws a cordon around the guerrilla area and then closes the ring.

(2) **Surprise attack**—By use of special troops and helicopters he will try sudden strikes which ensure that the guerrilla is kept moving if nothing else.

(3) **Trained bands**—He employs small but effective auxiliary police or army units who can move fast, know the country and the people and whose job it will be to destroy the columns, cut their communications and strangle their supply lines.

TECHNIQUES

The enemy will utilise surprise and will attack in bad weather, at night or when the roads appear impassable.

The enemy hopes to achieve this surprise by guarding his intentions and by constant change of direction. They also search constantly for guerrilla bases and comb the countryside as well as using patrols and reconnaissance units and local guides when on this job.

In following his basic anti-guerrilla techniques the enemy does the following:

(1) **Encirclement**—He cuts off all escape routes and then sets out systematically to destroy the guerrillas. Assembly points will be long distances from the guerrilla areas. The encircling lines will be fixed in accordance with the terrain. The aim is to close a ring around the columns and hold them.

Here the enemy's weakest point is his approach to the encircling areas from his assembly point. He will use vehicles and heavy weapons and strong reconnaissance units. When he has his ring completed he may use several methods to finish the job.

Some of these are:

(a) All encircling lines advance simultaneously towards the centre. This can only be done in relatively small areas. Otherwise great gaps occur in the circle.

(b) As one force digs-in the other lines close in. The idea is to drive the guerrillas against a wall of fire and kill them. The attackers keep pressing the guerrillas all the time.

(c) The wedge system—As the ring is completed other enemy forces drive strong wedges towards the guerrillas.

As the wedge splits up the guerrillas, several smaller rings are created. Each ring is then sealed off and the guerrillas caught within it are systematically destroyed.

Or if the enemy's intelligence tells him the guerrillas are operating from a special base, he will encircle the area and use shock troops to destroy the guerrillas.

(2) **Surprise attack**—The enemy avoids the trouble of encirclement and tries for a knock-out blow by a surprise raid.

He picks a time when the guerrillas are resting or least expect an attack. He reconnoitres beforehand to get the exact location of their base and its strength. And then he assembles secretly and as the guerrillas withdraw he moves up along the flanks.

(3) **Special bands**—When the guerrillas are weak the special bands go into operation. They track-down the guerrilla column, keep it on the move. They try to overtake, hold and then destroy the columns. Above all they go for the guerrilla H.Q.

For this purpose special police are used. They are constantly deployed to keep the guerrillas on the move and thus prevent harassing actions. They clear areas of guerrillas and try to wear them out. Mostly they are recruited locally and know the country well. Their forte is the surprise attack.

They watch supply areas and are specially armed for the job at hand. They are in constant radio touch with real units for immediate reinforcements.

THE ANSWER

The enemy's counter-guerrilla tactics sound formidable but they will be of little use against skilled columns who understand the motives behind them and stick to their original objective.

The guerrilla must know and understand that:

(1) No encirclement is complete enough that it can't be broken. There are always weak links and these must be found, probed and exploited. Note where units link-up.

(2) Proper intelligence can counter the special surprise blow. The guerrilla must always be on his guard.

(3) When the enemy attacks with his trained bands of hunting packs the guerrillas must fall on them and ruthlessly destroy them. Provide diversions. Hit the enemy's base. Cut him off. When one area is under enemy attack, increase activity in other areas.

The guerrilla must retain the initiative at all times. And he must stop the flow of information to the enemy. The guerrillas must be so trained that on capture despite the enemy's pressure they reveal nothing. For the enemy hopes to build up his knowledge of the guerrilla in this fashion: get his strength, its aims, its objectives, its leaders, its supplies, its bases and so on.

CHAPTER 10—GUERRILLA DEFENCE

Properly speaking a guerrilla never goes into a defence situation if he can avoid it. There are occasions, however, when the guerrilla will be unable to avoid a defensive battle.

Here is how he deals with such eventualities:

(1) **Encirclement**—The enemy attempts to destroy escape routes and then eliminate piecemeal the guerrilla columns.

(a) By reconnaissance the guerrilla probes the enemy circle.

He discovers the enemy's strong defensive lines then his weak points. Having discovered the weak spot—either in men or terrain—he stages a mass break-out.

(b) The enemy attempts to hide his movements from guerrilla observation. He proceeds from the assembly area so as to reach cordon areas at a definite time. His critical period is the approach from the assembly area into the cordon area. The guerrilla attempts a break-through at this time.

(c) Break up into smaller formations and either infiltrate the enemy lines (there are always unguarded spots between defense positions) or get out by use of still smaller formations. Assemble later.

(2) **Closing cordon**—All enemy encircling forces advance at the same time towards the centre. Terrain here helps the guerrilla for the enemy can only use this technique in small areas. The encircling lines cannot advance equally fast and big gaps occur. The guerrilla breaks out at these points.

(3) **Driving against wall of fire**—This consists of a stationary defence line with other forces driving the guerrillas back against it. The guerrilla probes the defence line and the terrain. If the area is not easily defensible he stages a break-through. If the line is too solid he escapes via the attackers who will inevitably be strung out and disorganised.

(4) **Driving in strong wedges**—The ring is held solid and strong wedges probe forward cutting the guerrillas into smaller bands and depriving them of freedom of action. They cannot find the weak links in the enemy.

After a time this situation leads to enemy confusion no matter how well he is organised. The guerrilla utilises the confusion and by the judicious use of scouts and snipers adds to it by striking various of the wedge forces. The guerrilla commander holds his main force for a break-out. By use of mobility and avoiding the wedges the guerrilla finds the enemy's weak spots.

(5) **Shock units**—The enemy encircles the area and uses a strong shock force to destroy the guerrillas.

The guerrilla avoids the shock-force and breaks through.

(6) **Surprise attack and hunt by special groups**—The answer of the guerrilla is always to guard against surprise, avoid hunting packs and manoeuvre rapidly.

ALL-ROUND DEFENCE

When taking up position however temporary even the smallest unit prepares for all-round defence.

In the case of a strong semi-permanent base preparation gets under way from the start for a sure-fire defence method.

The column takes up a triangular defence so that each section can support the other. Within the section the two-man teams and the section leader are similarly deployed.

Pickets and sentries are posted to cover all approaches. Grenade throwers and light infantry weapons cover ground which cannot be supported by the sections. Mines and anti-tank traps dominate all roads and paths. Mobile forces if possible should be prepared for enemy strikes.

If there is an encirclement of many columns, higher command organises the break-out. Diversionary attacks on the enemy's bases are organised from outside the ring. The enemy will be surprised in turn by flanking attacks designed to roll up his circle. Mass break-outs then occur.

If none of this is possible the columns break out on their own. If this is not possible the columns break up into sections and perhaps the sections into two-man teams and then get out.

The rule is: **Never give battle on the enemy's terms. Divert him by quick attacks in other areas. Hit him at his weakest point and drive a wedge through him.**

29

CHAPTER 11–GENERAL TECHNIQUES

The guerrilla must always remember that his main job is the destruction and break-down of enemy communications, administration and supplies, and not the capture of specific objectives. Therefore, the more the enemy is harried the better the result. The guerrilla can **always** harry the enemy by even small scale methods.

Trees felled across roads can cause long delays. Railway signal boxes can be effectively sabotaged. Telegraph and telephone lines can quite easily be put out of order.

Four men can fell 200–300 trees of one foot diameter in a day. These can be set up as barriers or barricades.

A stone tied to the end of a long piece of cord can be thrown over a wire and the wire lowered to the roadway and then cut.

When using explosives it should be remembered that a pocket knife, adhesive tape and matches (if possible, phosphorous matches) are indispensible.

HOW TO DO IT

Fit fuse in detonator at opening and crimp the det. to the fuse to stop it from slipping out. Always cut the other end at an angle to effect rapid lighting. Wind waterproof sticky tape around the connection point where the fuse fits the det. to prevent moisture.

Always see that explosives are in close contact with the object to be blown up—especially if it is metal. If contact is not possible increase the amount of explosive. Make sure each parcel of explosive is touching the next.

Ensure also that they are well tamped down as the success of any explosion depends on confining the charge in a restricted space. If using ammonal it is advised that the primer, fuse and det. be laid beneath the surface of the ground for safety.

Explosives actually sever only where they are in contact with the objective, except in a minor degree where this is due to blast effect.

For metal—Use a half pound of explosive for every half inch of thickness for a width of a foot.

For stone or brick—Use a half pound of explosive for a thickness of 10 inches by a width of one foot.

For woodwork—Use a half pound of explosive for five-inch thickness by a width of a foot.

It will be seen from these that the standard ratio of the above is 1-20-10. Further note, that should the object to be exploded be of circular form, calculate as though it were a square and use approximately 4/5ths of this amount.

For these three types of objects the following explosives can be used but make certain that a primer is also used with TNT, gun-cotton, gelignite and plastic.

One useful method of calculating quantities is to divide the thickness of the object in inches by 10 in the case of wooden targets, 20 in the case of stone and 1 in the case of metal. Then obtain the square of the figure and allow twice the amount for every foot of width. This will give the minimum effective quantity of explosives in pounds that is required. Always be on the generous side.

EXAMPLES

Here are some examples:

(a) To demolish a metal structure four inches thick by three feet wide.

$$4 \times 4 = 16.$$

Therefore 32 pounds of explosive are required for each foot. For three feet 96 pounds will be needed.

(b) To demolish stonework 10 feet in diameter:

10 feet = 120 inches.

This, divided by ratio of 20 is equal to six.

$$6 \times 6 = 36.$$

31

Therefore 72 pounds will be needed for every foot of width. As this is a circular object, allow the same width as thickness which gives a width of ten feet.

Thus explosive required for a square objective 10 ft. by 10 ft. is 72 x 10 or 720 pounds. As this is a circular section all that is required is 4/5ths of this amount or 576 pounds.

Roadways–Sixty pounds of explosive buried six inches beneath the level will cause a crater of 12-14 feet across.

When using explosives for mining or cratering, always tamp them tight. For this purpose it is preferable to use ammonal or gelignite without the aid of a primer.

Buildings–Sixty pounds for every 100 square feet of space on ground floor level is sufficient to demolish a two-storey house. It is necessary to close all doors and windows to get the maximum effect.

The best results are obtained when the charges are placed in the middle of each room.

Bangalore torpedo–This is one of the most effective methods of clearing barbed wire entanglements, and is prepared by use of metal pipe filled with explosives. Any pipe of an inch and a half or more diameter will do.

Make certain that the length of the pipe is never less than the depth or width of the entanglement, as it is the effect of the disintegration of the metal of the pipe which bursts asunder under the wire barricade, and not the explosive, that counts. Always ensure the pipe is kept at least 18 inches above ground, to keep the maximum tearing effect.

TO MAKE

To make Bangalore: Fill pipe with explosive and tamp well down. Lay primer and det. with fuse fitted at one end. Plug both ends with wooden stoppers, one of which should have a hole to allow the fuse to enter.

Should the torpedo be more than eight feet in length always

fit a fuse of F.I.D. through the middle of the explosive from end to end. This enables a perfect detonation to take place.

The torpedo is also effective against road or tank blocks, as it has a great disintegrating effect, and if arranged on small road blocks, are useful in dealing with enemy tanks.

Safety with explosives—**Always** allow sufficient length of fuse so as to get to safety before explosives goes up.

Clear the area of friendly personnel. **Never** attempt to examine enemy explosive charges. Leave it to the specialist. **Never** use naked lights or allow smoking near explosives. Do not use steel or iron instruments in handling, laying or tamping.

MAIN EXPLOSIVES

Gelignite is brown, used for cutting. Initiate with primer. Can be initiated with bullet. Keep in cool place.

808 looks like shaving soap. Yellow colour. Cutting or cratering. Do not inhale fumes.

TNT is yellow thready cake. Use for cutting and is liable to dry up and crumble. Store in box until needed.

Ammonal has greyish dusty appearance. For cratering or excavating. Keep away from damp.

Wet gun cotton looks like white candle wax cake. Used for cutting. Liable to dry and crumble. Store in box.

Plastic looks like putty. Used for cutting or excavating.

822 is a liquid for general purpose use. Must be initiated by det. fuse and a primer is required.

Fuses are either burning or det. type. Check rate of burning for speed. All burning fuses have a blackish powdery core.

DET. TYPES

Cordtex—white and silver core. Damaged by damp. Allow more. Burns at 6,000 yards per second. **FID**—yellow core with leaded cover. 5,000 yards per second. Operates under water. Make sure no bends or kinks. Throw away first foot when cutting off a portion. **Primacord** has yellow cord and coarse yellow cover. 6,000 yards per second. Allow extra two feet.

SABOTAGE

Put sugar or sand in petrol tanks. To immobilise enemy aircraft best place is the elevator. This is made of fragile material and can be easily damaged. If the elevator is out of action plane can't rise into the air.

When attacking enemy railway communications try to blow up a bridge since this is the most difficult to repair. Blow up a train so as to block the track. This will distort the rails and prevent enemy traffic passing for quite some time.

Unbolt railway lines at an embankment or curve. This tilts the train off the lines for a considerable time.

If there is not time for this a half hundredweight of fat, lard or grease, spread on an upward gradient will prevent the engine gripping the rails.

ANTI-AIRCRAFT

Guerrillas on the move must always be on the watch for enemy aircraft. Also when resting.

One of the best protections is a standard warning system. The columns should always be dispersed when moving in daylight in case of aircraft. Air guards should see and watch for aircraft when in camp. The drill is—scatter, hit the ground, take cover.

Slit trenches and dispersal will be sufficient protection against enemy bombing of a camp.

HELICOPTERS

Enemy increasingly uses helicopters against guerrillas. The Westland Whirlwind is for troop assembly and dispersal. The Two-Seater Saunders Roe Skeeter is for staff officers and reconnaissance.

CHAPTER 12–BATTLE NOTES

The first essential is to kill the enemy without being killed yourself. The second is take advantage of all cover. Equipment must be camouflaged. This goes for face and hands too. At night silence in movement is vital. When sniping in open country never operate from same spot twice. Continually being on the move is important for guerrillas. The enemy is thrown off by this and overestimates numbers and power. This makes him sometimes alter his plans.

RAIDS

The plan should be simple and the attack timed. Rehearse if possible. Special training may be required depending on the job at hand. Intelligence must be excellent for this planning depends.

AMBUSH

Preparations must be well camouflaged and groups should move to assembly point in small formations. No contact should be made with local population. All fighters must be accurately briefed on plan, withdrawal and later assembly point.

Can be turned into battle drill in this fashion:

1. The fire base—main concentration of fire and opens up on enemy first.

2. The assault party—moves in for the kill if the situation requires and if not covers withdrawal of fire base.

3. Cut off—seals enemy retreat and covers assault. Groups will not be of equal size. The fire base will have heavier concentration of fire.

Site—very important. Must have cover and sheltered line of withdrawal. Fire base must be able to cover battle area and allow no cover to enemy who will try to encircle attackers.

Road may be mined to hit first enemy truck, and last truck may also be trapped by similar mine to rear—blown not by contact but by electrical connection. Charges can be buried at a depth of nine inches or a foot. Four pounds per foot in length. These are enough to smash a lorry.

Signals must be clear and understood and ambushers must be in contact. Get-away must be swift. Look-outs are needed before enemy approaches. Aiming marks are valuable in killing area and can be prepared beforehand.

STRONGPOINT

Any enemy post is a strongpoint. Use fire base and assault section. Isolate post, use surprise to get to objective, lay charge to gain entry. First section moves in and clears ground floor. Section section clears, top, blow up, withdraw to prepared assembly point.

Rehearse if possible and each man should know his job.

FORCED MARCH

Move at night, change direction frequently, avoid inhabited localities and towns, go over heavily guarded territory in twos and threes, stay clear of roads. Check an assembly point and stopping points along the way. Guard front and flanks.

PATROLS

Mission must be precise. Men must know what their job is. A get-away man should be provided.

Select routes forward, stopping places, assembly, point, withdrawal. Study terrain, weather conditions, enemy dispositions. Use scouts.

MOVEMENTS

Use of formations—for open country use diamond. Use

Indian file for close country. Be in position for all-round observation, each member has special area to observe.

SEARCH BUILDINGS

Cover front and rear. The two men advance cautiously and enter by blind side. Have signals.

Crossing roads—At bend. All together or in groups.

CONDUCT OF AN ATTACK

1. Select Forming Up Point (FUP).

2. Issue orders and if need be make last minute reconnaissance.

3. Stay under cover near objective until preparation is complete. This should be as brief as possible—matter of minutes or column will lose element of surprise.

4. Basis of attack is by fire and movement. First base covers movement of assaulting groups. Hold fire until last minute or until prearranged signal.

5. The assault is put in with aggressiveness and efficiency, each man knowing his job and doing it. Mission is then completed.

6. Withdrawal is also by fire and movement.

7. Break contact with enemy and reassemble at assembly point which is prearranged. Alternative assembly point is also necessary.

8. Get back to bases.

Orders—Should be brief and cover following points.

1. **Information** on the enemy.

2. **Mission**—what we have to do.

3. **Method**—how we will carry out mission.

This covers:

    (a) The plan

    (b) The route forward

(c) Formations
(d) What each section will do at objective.
What each man will do. (This latter in section leader's orders.)
(e) Withdrawal
(f) Assembly point
(g) Alternative assembly point
(h) How to get back to bases.

4. **Administration**—includes (a) weapons distribution; (b) equipment; (c) where extra ammo is; (d) first aid and wounded.

5. **Communication**—signals used for attack, withdrawal, and others.

Watches should be synchronized and check back on orders and ask for questions.

In attacking enemy strongpoints maximum use should be made of explosives and the attack should be made silently at night. When covering fire opens up it should be intensive and be directed at enemy loop-holes and if possible be on (for LMGs) prepared fixed lines (use of two sticks to left and right).

The assaulting groups should be at as wide an angle as possible from fire group (90 degrees is ideal but may not always be possible). When entrance has been made the assault force should move in rapidly according to prepared drill, cover one another's inside movements, clear ground floor rooms by grenades and sub MG fire.

Second assault group should come, if possible, from top down. Have signals prepared. First group could then get back out and cover enemy driven downward.

Demolition group then destroys enemy base.

ESTIMATE OF SITUATION

For the guerrilla commander it is all important that he should have a correct estimate of the situation in all operations. In this way he discovers weaknesses in his plan, what extra

38

information he requires, what special type of equipment he may have to use in a particular attack.

The estimate is an analysis of a set of circumstances to determine the best course of action to pursue. Its form is:

1. **The mission**—what is the task before him.

2. **Situation and courses of action**—(a) Enemy dispositions, activities, strength, reinforcements, time and weather, his own equipment, supply and transportation. Local politics of population. (b) Enemy capabilities—What the enemy is capable of doing to stop the job being carried out. What he is likely to do—or the courses of action open to him. (c) Our own courses of action—the various methods by which the mission may be carried out.

3. **Analysis of opposing courses of action**—Determine the effect of each enemy capability on our own courses of action.

4. **Compare our various courses of action**—Weigh the advantages and disadvantages. Decide which course of action promises to be the most successful in accomplishing the mission.

5. **Decision**—Translate the course of action selected into a concise statement of what you will do—also when, where, how you will do it.

INTELLIGENCE

Top intelligence work is necessary for guerrilla operations. Intelligence is the collection, analysis, distribution of enemy information. It may be collected many ways. It must be evaluated and analysed—check one against the other. Then it must be passed along to higher and lower command.

Information gathering is a continuous operation. It will be done by reconnaissance, by patrols, by establishing observation posts and listening posts which note the enemy's movements and routine, from the local population, by agents and so on.

The intelligence officer or commander should have an intelligence plan. This consists of (1) **Essential elements of infor-**

mation—or what particular information is required about the enemy. (2) **The analysis of EEI**—the IO visualises what the enemy would do if he intended exercising any of the capabilities raised in EEI. For example, in his security measures of a barracks has he installed an alarm system, has he reinforced the defences, has he put more barbed wire on the walls, etc.? The IO looks for signs which will answer the specific questions raised in the Essential Elements of Information. There are indications that the enemy intends to do this or that. (3) IO lists reconnaissance and other intelligence methods to develop indications in column two. Is the enemy preparing for an attack on guerrillas? What are the indications that he is? Check these by specific missions. (4) Agencies to be employed in seeking information—patrols, OPs, listening posts, local inhabitants, agents and so on. (5) Hour and destination at which information is to be reported.

INTELLIGENCE PLAN

The IO works according to an intelligence plan. This is for his own guidance. But if specific information is required on an enemy target then he will make an **intelligence summary** on that target. That is he gives a full intelligence estimate (see estimate of the situation) on enemy-strength, disposition, defences, reinforcements, weather, terrain, local politics and so on. Here he will list everything pertaining to that enemy target.

The IO must know everything about the enemy. If he doesn't he must take steps to get his information. Then he must pass it along without delay.

There is only one sure way of getting information on the enemy: Go out and look for it.

A complete intelligence network should be built up in the enemy occupied area. Agents should be located in important localities and specialists should be sent after key information.

As many forces as possible should be used in intelligence

work and the collection should be so organised that the agents are not put in jeopardy. Couriers would collect and agents would be walled off from contact with any other guerrillas.

Such a chain of information, difficult to build up, would keep the guerrillas informed of everything happening in enemy-controlled territory.

Counter-intelligence is also an important activity of guerrilla IOs. It means guarding our own security, denying information to the enemy and tracking down enemy agents.